AF575922

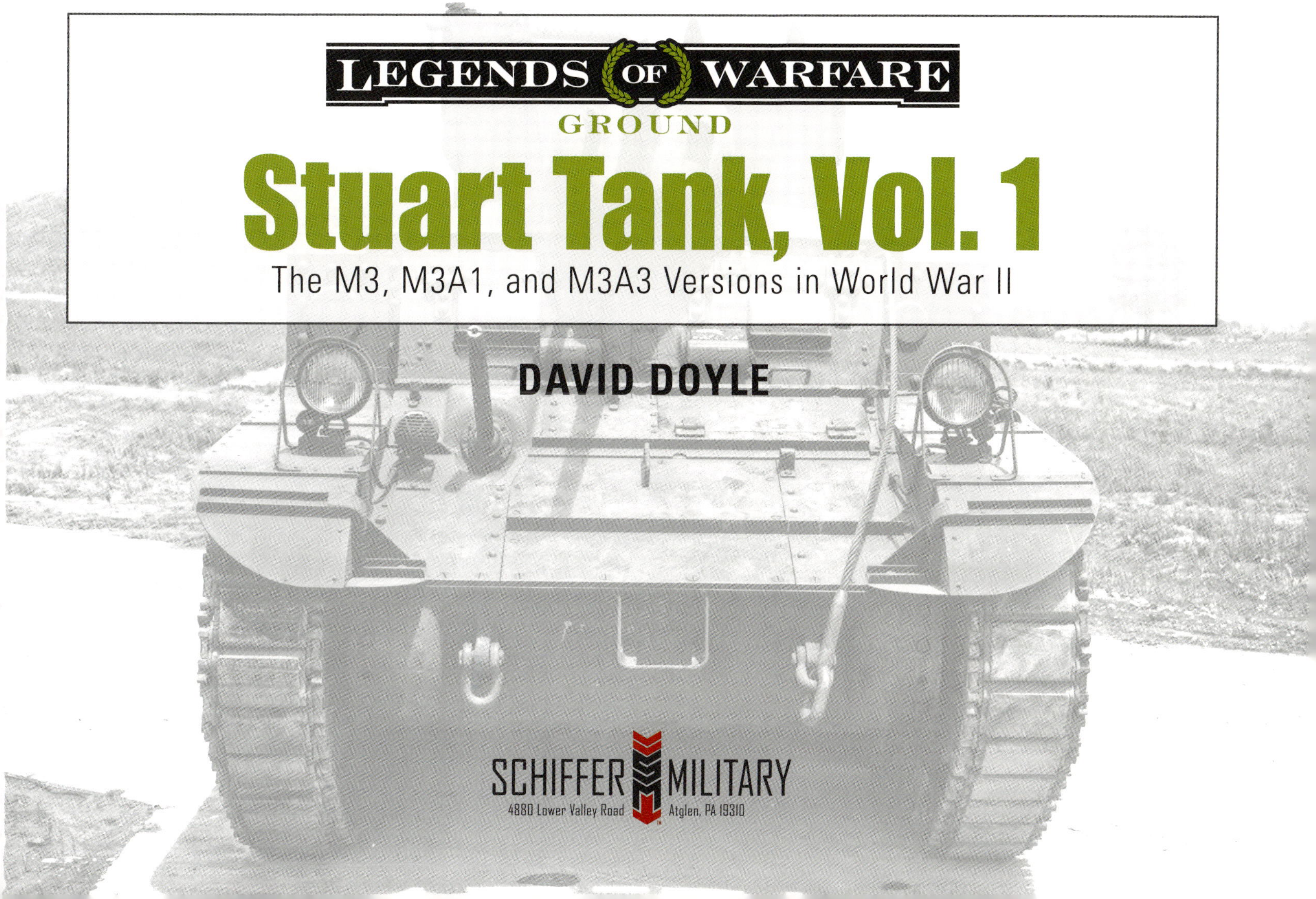

Legends of Warfare

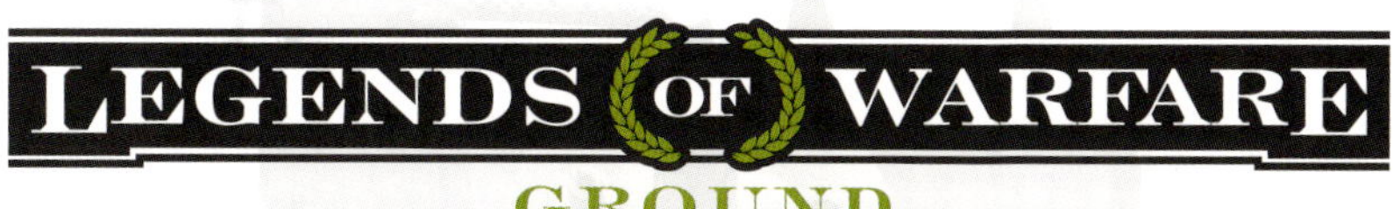

Ground

Stuart Tank, Vol. 1

The M3, M3A1, and M3A3 Versions in World War II

DAVID DOYLE

SCHIFFER MILITARY

4880 Lower Valley Road Atglen, PA 19310

Library of Congress Control Number: 2018956656

Designed by Justin Watkinson
Type set in Impact/Minion Pro/Univers LT Std

ISBN: 978-0-7643-5660-5
Printed in China

Published by Schiffer Publishing, Ltd.
4880 Lower Valley Road
Atglen, PA 19310
Phone: (610) 593-1777; Fax: (610) 593-2002
E-mail: Info@schifferbooks.com
www.schifferbooks.com

Acknowledgments

As with all of my projects, this book would not have been possible without the generous help of many friends. Instrumental to the completion of this book were Tom Kailbourn, Rob Ervin, Steve Greenberg, Skip Warvel, the late Fred Ropkey, Scott Taylor, Kurt Laughlin, Central Susquehanna Community Foundation, Jim Gilmore, Joe Demarco, the staff and volunteers at the Patton Museum, and the staff at the National Archives. Most importantly, I am blessed to have the help and support of my wife, Denise, for which I am eternally grateful.

All photos are from the collection of the US National Archives and Records Administration, unless otherwise noted.

Contents

Introduction

The United States gained useful experience in armored warfare during the First World War. In spite of the brevity of its use, the fledgling tank army planted the seeds of its future use with such notable individuals as George S. Patton, who commanded a tank unit in that conflict. The rush to disarm, although well-intentioned, relegated the control of tanks to the chief of infantry and eliminated the Tank Corps through the National Defense Act of 1920.

US tank design followed the concepts set forth from the war—either large, such as the "Mark" series of British tanks, or small, such as the Renault FT-17. Given the severe budget constraints on the postwar army, American tank design more closely followed the latter pattern. In fact, Van Dorn Iron Works, the Maxwell Motor Co., and the C.L. Best Co., built licensed copies of the FT-17. Known as the M1917 6-Ton tank, it had been intended to equip US armored units in the final years of the war, had it continued. Due to lack of funds, only about thirty-five new tanks were built in the United States from 1920 until 1935, and these all were simply prototypes of various concepts.

In the mid-1930s, development began on a new series of light tanks that abandoned the negative concepts and renewed the positive ones set forth in World War I. What became known as the M3 and M5 series of light tanks was born from these vehicles, identified by their "T" prefix. The first vehicle developed with the characteristic drivetrain layout and vertical-volute-spring suspension system was the T5. Introduced in 1934, it was powered by a Continental Radial R-670 gasoline engine. This robust and dependable engine could develop 235 horsepower. The T5 was armed with a .50- and a .30-caliber machine gun in a pair of turrets and an additional .30-caliber machine gun mounted in the right front bow plate. It stuck to the original weight class of the World War I light tanks, coming in at approximately 6 tons. Testing at Aberdeen Proving Ground, Maryland, in the summer of 1935, resulted in the suggestion that the T5 use a fixed casemate mounting four .30-caliber machine guns. The T5 was then converted at the Rock Island Arsenal, becoming the T5E1.

A similar vehicle was also under development in the mid-1930s. Known as the T2, it was initially equipped with a suspension very similar to the British Vickers 6-ton tank and a single turret. It progressed through a series of modifications and testing in 1934 and was later standardized as Light Tank M2A1. Nine were built in 1935. Light Tank M2A2, a twin-turreted variant, was standardized at the same time.

The twin turrets of the M2A2 were armed with a .50-caliber machine gun on the left and a .30-caliber on the right. A total of 237 M2A2s were built, making it the most numerous tank produced in the late 1930s. Its use in training exercises proved highly useful for the development of new armored tactics. Crews and maintenance personnel also gained field experience that would prove invaluable in later years.

A further development of the M2 design was the M2A3. The hull of the vehicle was lengthened, increasing the contact area of the tracks by 11 inches and lowering ground pressure to increase mobility. The M2A3 went into production in 1938. A total of seventy-three were produced. The many changes in the basic design had now increased the tank's weight to 11 tons. Automotive performance remained high, with the vehicle reaching a top speed of 37.5 mph.

The cavalry had also been excluded from possessing tanks due to the Defense Act of 1920, so parallel tank development within that branch used the term "combat car." These vehicles were primarily intended for the role of reconnaissance. Developed concurrently with the M2 light tanks, the M1 Combat Car shared many components, such as the engine, transmission, tracks, and suspension. They were single-turreted vehicles mounting a .50-caliber and 30-caliber machine gun side by side. An additional .30-caliber was mounted in the bow plate. The M1 had a top speed of 45 mph, and total production had reached approximately ninety units by 1937.

Production of the M1 and M2

Because of the stagnation of medium-tank development, the M1 Combat Cars and M2 series Light Tanks formed the backbone of the fledgling US armored divisions. Their numbers greatly exceeded the number of the available medium tanks through the late 1930s.

Rock Island Arsenal began turning out the twin-turreted M2A2 in 1935, and 237 vehicles were produced by the end of 1937. It was the first true US production tank since the end of World War I. The tanks were initially made with rounded turrets, but these were later replaced with turrets of flat plate construction. The M2A2 remained similar to the T5 in layout, with its Continental Radial W-670 gasoline engine and vertical-volute suspension.

To improve engine access and reduce ground pressure, the hull length of the M2A2 was increased, becoming standardized in 1938 as the M2A3. Seventy-three of these tanks were produced in 1938. Armor thickness was equal to its foreign contemporaries at ⅞ of an inch or 22 mm, and its combat weight was 10.5 tons. The M2A3 was able to reach 37 mph on good roads.

When the Spanish Civil War broke out, proxies for the opposing sides used the conflict as a testing ground for their latest tank designs. Vehicles such as the Soviet T-26 and the German Panzer II demonstrated the need for tank armament beyond that of machine guns. Development quickly ensued to arm the M2 series with a 37 mm antitank gun—a weapon equaling or bettering those then in foreign use. The newly armed tank was standardized as M2A4 in late 1939, and it went into full production in May 1940.

In addition to the main gun, the M2A4 mounted five .30-caliber machine guns. Riveted flat armor plates formed the turret. The side armor plates had a total of seven shuttered-vision doors. A raised cupola with vision slots for the commander adorned the top of the turret. The turret lacked a basket, making it difficult to move within the rotating turret, and the tall center drive-shaft tunnel presented a significant obstacle to the crew. The upper hull of the M2A4 was welded, while the lower hull remained riveted.

The majority of M2A4s had the seven-cylinder W-670 Gasoline Continental Radial engine, which produced 250 horsepower. A small portion of production used the Guiberson Radial Diesel.

Armor protection was quite good for its day, with a maximum of 1.0 inch. The top speed of the M2A4 remained 36 mph, even though the weight had increased to nearly 13 tons. The standard vertical-volute suspension and dual-pin, rubber-padded tracks of its predecessors remained.

Nearly all 375 M2A4s were produced at American Car and Foundry. Combat use was brief, with a few seeing action with the USMC on Guadalcanal. Thirty-six were sent to Great Britain under the Lend-Lease Act.

Development of the Combat Car family progressed concurrent with that of the M2 Light Tank series, and they continued to share a common suspension, tracks, and drivetrain. However, the .50- and .30-caliber machine guns were mounted in a single turret. Initially, this was D shaped with a rounded rear section, but it later gave way to one of flat armored plate construction.

The M2 Combat Car was light and fast—the M1 version weighed in at only 9.5 tons. They continued to meet the reconnaissance needs of the cavalry divisions throughout the prewar period. Combat Cars were the first of the Stuart family to utilize the trailing-idler suspension that became common on the M3 and M5 light tanks.

Standardized in 1938, the Light Tank M2A3 was a twin-turreted vehicle with a longer hull than the M2A2. The sprocket and bogie assemblies were similar to the forthcoming M3-series light tanks, but the idler was mounted higher. This M2A3 is shown at the Rock Island Arsenal on September 8, 1938. *Rock Island Arsenal Museum*

The next development in US light tanks was the M2A4, which had the same hull as the M3A3 but replaced the two machine gun turrets with a single turret armed with a 37 mm cannon and a coaxial .30-caliber machine gun. The turret of the pilot M2A4, seen here at Aberdeen Proving Ground, Maryland, on May 11, 1939, lacked a cupola, a feature that would be included on production M2A4s. There was a bow-mounted, flexible .30-caliber machine gun, and a fixed .30-caliber machine gun was installed in each sponson to fire forward: these features would carry over to the early-production Light Tank M3. *TACOM LCMC History Office*

While the Army was developing the various models of Light Tank M2, parallel series of vehicles, the Combat Car M1 and Combat Car M2, also were being readied as tracked, armored scouting vehicles. Shown here is a Combat Car M1, whitewashed for winter camouflage purposes. The turret included two machine guns: a .30-caliber on the right and a .50-caliber on the left. Also, there was a flexible .30-caliber bow machine gun. *Patton Museum*

The Light Tank M2A2 was similar to the Light Tank M2A3 but with a shorter hull. Here, mechanics use a boom on a truck to remove a Continental R-670 radial engine from an M2A2, registration number W-30314. *Jim Gilmore collection*

A number of Light Tanks M2A3 and M2A4 are loaded on railroad flatcars for shipment cross-country. The first two tanks are M2A3s, while the next six are M2A4s, followed by more M2A3s. *Patton Museum*

The United States transferred thirty-six Light Tanks M2A4 to Great Britain under the Lend-Lease program, and the British used these vehicles to train personnel destined for Stuart Light Tank crews. Here, British army mechanics are getting a newly received M2A4 nicknamed "Al Capone" ready for operational use. *Patton Museum*

CHAPTER 1
Light Tank M3

The first of the production Light Tanks M3 is shown during testing at Aberdeen Proving Ground, Maryland, on April 14, 1941. This tank was serial number 1 and Army registration number W-30978. The initial-production M3s were assembled by American Car and Foundry, Berwick, Pennsylvania, starting in March 1941, and a total of approximately 100 were completed. The turret and hull of the initial-production M3s were of riveted, face-hardened armor construction, with some welding on the hull. The turret's ordnance drawing number was D37812. *Jim Gilmore collection*

With the German army devouring much of western Europe in the spring of 1940, the Ordnance Department recommended a series of changes to the M2A4 to keep its design on par or ahead of its foreign contemporaries. These many changes resulted in the Light Tank M3.

The M3 now had increased protection, with 1.5 inches of frontal armor. A new cast section now covered the area around the transmission and final drives. Improvements were also made both to the turret and the gun mount. The new M22 combination mount housed the recoil mechanism within an armored tube, and a cast outer rotor shield or mantlet protected the front of the turret. Face-hardened armor plates of riveted construction were used on the turrets of initial production vehicles.

In order to reduce ground pressure, the suspension was modified with a load-bearing trailing-idler wheel. This increased the contact area of the track without lengthening the vehicle. Performance of the M3 continued to be good, in spite of its combat weight now rising to 14 tons.

American Car and Foundry began production of the M3 in March 1941. Several production batches resulted in a total of 5,811 vehicles manufactured through January 1943. This included 1,285 tanks equipped with Diesel engines.

As the size of the army exploded in 1941, the M3 helped to expand the newly formed American armored units. However, many vehicles were sent to training units unarmed, due to shortages of the main gun and mount.

The M3's range was restricted to only seventy miles because of its scant fuel capacity of just fifty-four gallons. This limitation would dog the M3 series throughout its production.

Light Tank D38976 M3 Production Turret

Significant problems were revealed by ballistic tests of the early, riveted M3 turret. Those tests showed that even nonpenetrating rounds could still deform the plating. Rivets, broken loose by impact, could also become projectiles inside the tank.

A new turret was designed incorporating welded, face-hardened armored plates that mimicked the shape of the older design. Its ordnance drawing number can identify this turret: D38976. The heat of welding did not drastically affect the performance of the nearby armor, and the lack of the internal framework also saved some weight.

Vision slots in the cupola in the turret were retained, as were the armored flaps covering the pistol ports. The welded turret was approved for production in December 1941. Hull construction remained unchanged. The new turret design was found on about 1,600 M3s.

The face of the new turret was attached with screws to allow for easy removal of the gun. The M22 gun mount was retained, and it allowed for 10 degrees of right and left movement. The gunner used a shoulder mount for precision aiming.

Although the M22 mount was retained throughout the production of the D38976 turret, the longer M6 37 mm gun replaced the M5 gun roughly halfway through.

Ammunition storage consisted of 103 rounds of 37 mm ammunition. An M51 Armor Piercing Capped round was available, and it had a muzzle velocity of 2,900 feet per second. It could penetrate over 2 inches of armor at 500 yards. High explosive and canister rounds were also provided.

Light Tank M3 D39273 Turret with Oval Cupola

The introduction of the new D39273 turret design marked the first major change in M3 production. Two pieces of rolled, homogeneous armor plate welded together at the rear now formed the side armor of the new turret. The same removable gun mount was retained. Armored doors that included protectoscopes now replaced the pistol ports. Protectoscopes incorporated an armored-glass periscope into the casting, which deflected small-arms fire away from the interior.

The multiple flat surfaces of the cupola were replaced with two rolled plates that created an oval shape. It was topped with two small hatch panels and armored vision slots. Oddly, a circular aperture was contained in one of the hatch plates for a periscope but was never utilized. Ballistic protection was greatly enhanced with the new design, and the reduction in welds sped up production.

Concurrent with the above changes, the earlier vision slots in the front of the hull were eliminated. The front hull doors were redesigned and equipped with protectoscopes. New sheet-metal boxes for storing track grousers and other items were added to the rear corners of the hull behind the air cleaners.

All these changes began with vehicle number 1946 at American Car and Foundry. This vehicle was later the subject of testing at Aberdeen Proving Ground in November 1941.

Both the Continental gasoline and Guiberson Diesel engines continued to be used in the production of the M3.

British Use

Before America's entry into World War II, one of the driving forces behind increased US tank production was the desperate need to supply British forces with capable vehicles. With the passage of the Lend-Lease Act in March 1941, priority was given to the shipping of US Light Tanks to the British. Brand-new M3s started arriving in Egypt in July of that year.

Modifications were made to the tanks on the basis of prior British experience in desert warfare. New sand shields, external stowage boxes, and fuel can racks were added. Several internal modifications were also made to the interior, which included the removal of the machine guns mounted in the sponsons.

The British had a habit of naming their Lend-Lease machines after American Civil War generals. The M3 Light tank was dubbed "Stuart." With British units, it served in the role of Cruiser tanks and proved to be very reliable—garnering the nickname "Honeys" from their crews.

Complaints included the lack of a turret floor, which made the tanks awkward to fight in. The crew had to step over the central drive-shaft tunnel and ammunition storage in order to traverse the turret. Steering the tank toward a target and then making final adjustments with the traverse of the gun mount solved this problem to an extent. This practice also kept the thicker frontal armor facing the enemy.

The long distances associated with desert warfare meant that its short range would limit the M3 Stuart. As more M3 Grants and M4 Shermans reached the battlefield in 1942, British armored units began regulating the M3 Stuart to the reconnaissance role.

The first Light Tank M3 is seen from the left side at Aberdeen on April 14, 1941. The suspension was the vertical-volute-spring type, with two bogie assemblies supporting two bogie wheels each. Three track-support rollers were attached to the side of the lower hull. Size 30 x 6-inch idlers, front-mounted sprockets, and T16 tracks completed the running gear. The idlers were a departure from those on the Light Tank M2, in that they were of a trailing design. The bottom of each idler was on the same level as the bottoms of the bogie wheels.

The driver's and assistant driver's front hatches are in the open position; each one is supported by a hold-open strut on each side. A step made of bent steel is on the front of the hull, below the glacis. On the glacis to the front of the assistant driver's position are a .30-caliber M1919A4 machine gun and a siren. A service headlight and its brush guard is on each fender.

As seen in an April 14, 1941, photo of the right side of the first Light Tank M3, an air filter was mounted to the rear of the sponson; a similar one was on the left side. A .30-caliber M1919A4 machine gun was on a fixed mount in the front of each sponson; the barrel of the right one is visible here, with a protective cover over it. The main gun was the 37 mm Gun M5 on the Mount M22, with a coaxial M1919A4 machine gun on the right side.

The rear of the first Light Tank M3 is shown. There were three pistol ports, hinged at the top, on the turret: one on the rear, and one on each side. Both of the air cleaners are in view, as are the pioneer tools on the engine deck and the rear of the upper hull. Engine-access doors are on the rear plate of the lower hull. An M1919A4 machine gun was on a flexible mount on the rear of the turret, for antiaircraft defense. Taillight assemblies are on each side of the rear of the upper hull.

Following approximately the one-hundredth Light Tank M3, an improved turret fabricated from welded, face-hardened armor was introduced, replacing the riveted turrets. This turret was designated by its ordnance drawing number, D38976. This tank, serial number 1366 and registration number W-301900, was photographed at Aberdeen Proving Ground on September 10, 1941.

The welded D38976 turret and cupola were similar in design to the earlier, riveted turret, and the hull was the same as the original riveted design. A vision slot for the vehicle commander's use was on each facet of the cupola, for a total of six.

The driver's and assistant driver's hatches are in the closed position on Light Tank M3, registration number W-301900. Each hatch had a vision slot, and there also was a similar vision slot on each side of the upper hull for the driver and the assistant driver. Protruding through the gun shield are the barrels of the 37 mm gun, the .30-caliber coaxial machine gun, and, on the left side, the aperture for the gunner's sight, with a semicircular rain guard over it.

A right-side view of Light Tank M3, registration number M-301900, shows the right-sponson machine gun barrel, below which is the assistant driver's side vision slot. Note that the bow machine gun has the early-type, slotted-barrel jacket, while the sponson machine gun has the later, perforated jacket. The turret armor was of rolled, face-hardened steel: 1½ inches thick on the front, 1 inch thick on the sides and rear, and ½ inch thick on the roof, with a 1½-inch-thick gun shield.

As seen in a photo of Light Tank M3, serial number 1366, with a welded turret, the rear of the hull is identical to that of the initial M3s with the riveted turret. On the right side of the engine deck to the front of the mattock head is a bracket for a radio antenna.

The roof and the cupola of welded turret on a Light Tank M3 are viewed from above just prior to a test in which these surfaces were subjected to .50-caliber armor-piercing and tracer fire from a 30-degree angle. The scene was Aberdeen Proving Ground, on July 7, 1941.

The same turret and cupola are viewed from the upper rear prior to being subjected to .50-caliber machine gun fire on July 7, 1941. Affixed to the rear of the turret is an experimental curved armor shield, to protect the pedestal of the antiaircraft machine gun; this shield evidently did not make it to production status.

Light Tank M3, serial number 1366 and registration number W-301900, is descending over a 3-foot-tall vertical wall during evaluations at Aberdeen Proving Ground on September 5, 1941. The officially rated maximum vertical wall for this tank was 24 inches.

In an effort to extend the rather limited range of the Light Tank M3 (the internal fuel capacity was 54 gallons), Aberdeen Proving Ground instituted Project No. 7-4-3 around late 1941. This involved mounting two jettisonable, self-sealing, 25-gallon fuel tanks on the top of the hull to the sides of the turret. These tanks were made of rubber-saturated fabric. This photo shows two of the fuel tanks installed on M3 serial number 727 on November 28, 1941. The fuel lines from the tanks were connected to special couplings on the fuel fillers on the top of the hull. The jettisonable fuel tanks went into production and frequently saw operational use.

Two crewmen in civilian clothes take a ride in a newly minted Light Tank M3. The sponson and bow machine guns have been dismounted, and a detachable windshield with a wiper assembly has been installed in the assistant driver's open hatch. The 1.5-inch thickness of the armor of the driver's and assistant driver's hatches is well illustrated. The driver also had a lower hatch, on the glacis, for ease of entering and exiting the vehicle. On the inside face of the commander's hatch door is a latch and a port with a sliding cover. *Central Susquehanna Community Foundation*

Parked outside an entrance to the American Car and Foundry plant (note the "A.C.F." sign above the door) is the 1,000th Light Tank M3 built. This tank had the D39876 welded turret with the small, plain pistol ports. A flaming-grenade symbol is painted on each side of the "1,000th TANK" inscription on the sponson. The vehicle seems to have been painted overall in white, or possibly in a light sand color. *Central Susquehanna Community Foundation*

The next step in the development of the Light Tank M3 was the introduction in late 1941 of a new turret, referred to by its Ordnance drawing number, D39273, fabricated from two pieces of rolled, homogeneous plate armor on the sides and rear, with a vertical weld seam joining them at the rear; a flat frontal plate; and a cupola with a football-shaped plan. Welded armor segments formed a flared bottom for the turret's sides and rear. Seen here with the new turret in a November 17, 1941, photo at Aberdeen Proving Ground is Light Tank M3, serial number 1946 and registration number W-302480.

Both the driver's and the assistant driver's stations retained the vision slot on the side of the upper hull. On each side of the upper hull, to the immediate rear of the air cleaner, was a new storage box made of sheet metal. These were for storing grousers and other gear.

The rolled, homogeneous-armor turret retained the flat frontal armor and the same gun shield as the preceding M3s. The turret now was provided with three pistol ports, hinged at the tops, each of which had a protectoscope: a bullet-resistant-glass periscope that eliminated the chance of small-arms fire entering a vision slot. The driver's hatch doors also were revamped and equipped with protectoscopes.

Light Tank M3, serial number 1946 and registration number W-302480, featuring the D39273 turret, is observed from the left side at Aberdeen Proving Ground on November 17, 1941. On the turret, around the sides and bottom of each protectoscope, was a beveled splash guard formed of weld beads.

As seen in a left rear view of the same Light Tank M3, the lids of the new storage boxes on the rear fenders were held down by straps attached to footman loops.

The M3 with the D39273 turret is viewed from the rear at Aberdeen on November 17, 1941. Jutting from the inboard side of the cupola are two hinges; the hatch door of the cupola was of a two-panel design. On the right rear of the turret is a second antenna bracket, in addition to the one on the engine deck.

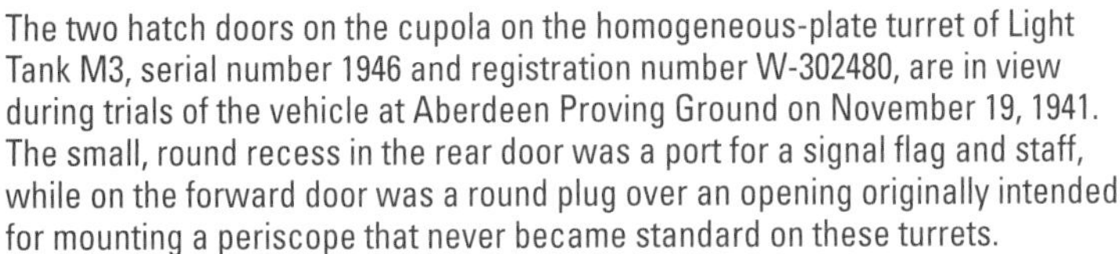

The two hatch doors on the cupola on the homogeneous-plate turret of Light Tank M3, serial number 1946 and registration number W-302480, are in view during trials of the vehicle at Aberdeen Proving Ground on November 19, 1941. The small, round recess in the rear door was a port for a signal flag and staff, while on the forward door was a round plug over an opening originally intended for mounting a periscope that never became standard on these turrets.

Light Tank M3, serial number 1946 and registration number W-302480, exhibits the revamped vision features on the driver's and assistant driver's hatch doors. On the outboard sides of the "buttoned up" doors are protectoscopes, while on the lower inboard corners of the doors are vision slots with swiveling, armored covers.

An American Car and Foundry photograph shows a new Light Tank M3 with no Army registration number painted on the sponson, with the rolled, homogeneous-armor turret. The factory caption on the photo refers to the turret as the "formed homogeneous turret." The driver had, in addition to his front hatch door, a door at the upper left of the glacis. The driver and the assistant driver entered and exited the vehicle through these doors, which was an extremely difficult process. In addition, the assistant driver had to scramble over the transmission and the drive-shaft tunnel to get to and from his seat, making his job even harder.

The same unmarked Light Tank M3 seen in the preceding photo is viewed from the right rear. This vehicle lacked the storage boxes on the rear fenders. The cupola on this M3 has the four vision slots that were the norm on the rolled, homogeneous-plate turrets: three on the right facet of the cupola, and one on the left facet. *Central Susquehanna Community Foundation*

A Light Tank M3 with the D39273 turret and vision slots in the cupola has paused during a road test. No Army registration number is visible on the sponson, so this probably was a tank under evaluation by the manufacturer. A tow cable is attached to the front-right tow shackle.
Central Susquehanna Community Foundation

On outdoor display at the 1st Cavalry Museum, Ft. Hood, Texas, is this Light Tank M3 with a D39273 turret. Round, steel plugs have been welded over the openings for the sponson machine guns. No headlights or brush guards are present on the front fenders, but there is a blackout marker lamp on each fender. Two parallel storage racks for grousers are on the side of the sponson. *Photo by author*

The 1st Cavalry Museum's M3 as viewed from the rear. *Photo by author*

The antiaircraft machine gun pedestal has been removed from the rear of the turret, and the holes for its mounting screws are visible. To the lower right, ajar, is the armored cover for the left fuel filler. To the right of the armored cover is the ventilating air grille for the engine. *Photo by author*

The weld seams on the front edge of the cupola and along the bottom edge where it is joined to the top of the turret are visible in this close-up photo. Also present are the left vision slot in the cupola and the left pistol port with a slot for the objective of a protectoscope. *Photo by author*

During combat, a projectile strike on riveted armor could cause rivets to shear off and ricochet inside the vehicle, causing death or injury to the crew and damage to the tank. That is the main reason welded turrets were introduced to the Light Tank M3. By late 1941, the decision was made to weld the hulls of M3s. An example, registration number W-303034, is viewed from the upper left in this photo. This type of welded hull would become standard on the Light Tank M3A1.
Central Susquehanna Community Foundation

Light Tank M3, serial number 3213 and registration number W-306565, photographed at Aberdeen Proving Ground on January 9, 1942, had the welded hull. A small detail concerning the sponsons of the welded hull was that the front and rear vertical edges of the side plate were butted to the small, laterally oriented front and rear plates. By contrast, on the earlier, riveted hulls, only the front edge of the side plate was thus butted: the rear edge of that plate was flush with the rear face of the rear plate of the sponson.

As seen in a right-side photo of Light Tank M3, serial number 3213 and registration number W-306565, at Aberdeen on January 9, 1942, the sides of the lower hull of the welded-hull tanks lacked the vertical joint near the center of the riveted-hull M3s. However, a feature new to the welded hull was the over-and-under hook that was fastened to the lower hull between the second and third bogie wheels.

The top and rear plates of the upper hull were fastened in place by countersunk slotted, oval-head screws.

On the rear of the upper hull of Light Tank M3, serial number 3213 and registration number W-306565, are two brackets and two footman loops for holding a shovel. On each side of the rear of the upper hull, to the rears of the storage boxes, are taillight assemblies. The retainer straps for the lids of the storage boxes appear to have been leather and were definitely not of webbing material.

This welded-hull Light Tank M3 is on outdoor display at the 3rd Armored Cavalry Museum, Ft. Hood, Texas. Round plugs with beveled layers of weld beads surrounding them are over the openings for the sponson machine guns. These plugs sometimes were installed during World War II when the sponson guns, always of limited utility as well as consumers of space, were dismounted. *Photo by author*

The storage boxes and the taillights are missing from the rear fender of the welded-hull M3 at the 3rd Armored Cavalry Museum. *Photo by author*

Details of the right sprocket assembly are shown. Each sprocket is attached to the drum with fourteen hex screws. Some of the end connectors of the track also are in view. *Photo by author*

The top of a bogie assembly is shown, including the track skid. The cylindrical structures inside the bogie bracket are the vertical volute springs, which buffer and help support the suspension arms and bogie wheels. *Photo by author*

A complete bogie assembly, the front-right one, is illustrated. At the top of the bogie bracket is an oblong track skid. On the bottom of the bogie bracket are the suspension arms. The rubber tires of the bogie wheels of the M3 at the 3rd Armored Cavalry Museum are in various states of deterioration. The track-support rollers are mounted on the hull, separately from the bogie assemblies. The right over-and-under twin-pronged hook, a feature introduced with the welded-hull Light Tank M3, is on the hull to the rear of the second bogie wheel. *Photo by author*

A close-up view of the front corner of the right sponson, the plug for the machine gun aperture, and the assistant driver's side vision slot and hatch door also provides details of the flared lower part of the turret, formed of welded segments of armor. *Photo by author*

The M3 at the 3rd Cavalry Museum is equipped with T36E6 double-pin steel tracks with parallel grousers on the treads. These tracks, more commonly associated with the Light Tank M5A1 and the 75 mm Gun Motor Carriage M8, nevertheless were approved and available for use on late-production Light Tanks M3. *Photo by author*

The engine deck and the rear of the turret of the 3rd Cavalry Museum's Light Tank M3 is displayed. Footman loops for strapping down equipment are distributed around the deck. In the foreground, just beyond the row of slotted screws, are countersunk holes for screwing down the engine deck. To the left, an air intake pipe curves down from the left air cleaner into the engine compartment. *Photo by author*

In a view looking down at the left front of the engine deck, at the bottom center is the armored cover for the left fuel filler, next to which is one of two cradles for holding an auxiliary fuel tank. (The other cradle is partially in the shadow to the top right.) These cradles as well as hooks and latches for the straps holding the tank in place are attached to a steel plate, which is perforated to provide clearance for the rivet heads on the upper hull. *Photo by author*

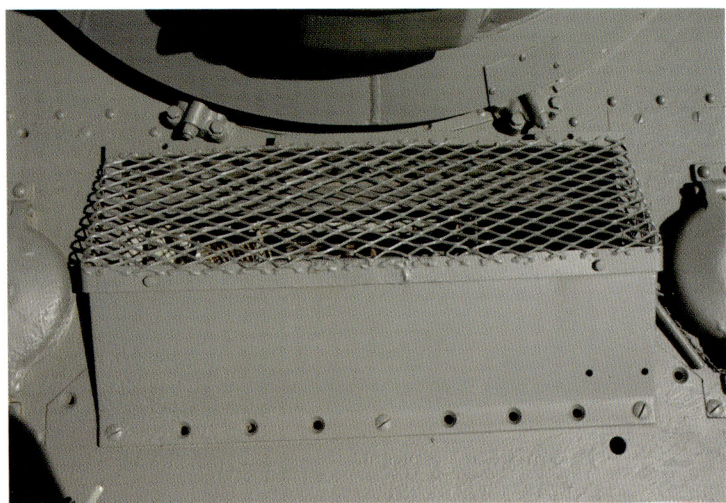

The engine air intake on the front of the engine deck consisted of a metal hood on the rear and sides and an expanded-steel-mesh grille on the front. Only three of the original nine slotted, oval-head screws are present on the flange on the rear of the intake. The two fittings to the front of the grille and below the flared lower part of the turret are two of the eight turret bearing and roller brackets. *Photo by author*

The right air cleaner and air hose are seen from above. The two air cleaners were the oil-bath type. The plate to the upper right with the grab handle welded on is the cover for the battery box. *Photo by author*

The cupola and hatch doors of the Light Tank M3 with the rolled, homogeneous-armor D39273 turret are viewed close-up. Instead of the round plate of relatively thick steel that normally was screwed down over the unused opening for the periscope on the front door of the hatch, a thin metal cover is attached without screws. On the rear of the cupola is the pedestal for an antiaircraft .30-caliber machine gun. *Photo by author*

The vision slot in the front of the cupola is shown, as are the two hinges for the front hatch door. Note the quarter-round weld bead between the turret roof and the cupola. *Photo by author*

The gun shield of the 3rd Cavalry Museum at Ft. Hood, Texas, has a plug over the opening on the left side for the gunner's sight. *Photo by author*

In a view taken above the center front of the turret roof, the gun shield and part of the 37 mm gun barrel are shown. The gun shield of the Combination Gun Mount M22 was 1.5-inch armor. The slotted screw recessed into the protrusion at the front of the turret, behind the gun shield, was for the top trunnion. This trunnion, along with a bottom one, were part of a mechanism that allowed the gunner to fine-tune the traverse of the gun 10 degrees to either side of the center, using pressure applied to a shoulder rest attached to the gun mount. *Photo by author*

On the right side of the gun shield is the aperture for the .30-caliber coaxial machine gun. To the lower left is the front end of the flared bottom of the turret. *Photo by author*

Light Tank M3, serial number 4927 and W-308360, photographed during evaluations at the Ordnance Operation, General Motors Proving Ground, on July 4, 1942, was a late-production, welded-hull M3 equipped with a new rolled, homogeneous-armor turret that lacked the cupola and had two hatches in the rear of the roof. This turret was designated the D58101. The tank also had a new combination gun mount, the M23. Quantities of these tanks were sent to the British, who named them the Stuart Hybrid. The basic turret shape would soon be used on a new model of Stuart: the Light Tank M3A1.

CHAPTER 2

Light Tank M3A1

Various improvements to the M3 Light Tank series began to be explored during 1941. Many of these new features were incorporated into production and were later standardized as the M3A1 Light Tank. During May 1942, the new tank began replacing the M3 in production.

Modifications to the D58101 turret included eliminating the cupola and replacing it with a pair of hatches in the rear turret roof. Periscopes were now provided for the commander and gunner. The front armor plate was now welded to the turret and equipped with a basket and floor, which moved with the turret during traverse. Power traverse and gyrostabilization of the gun were also introduced. Gyrostabilization was a significant improvement that greatly increased the chances of an accurate shot with the main gun while the tank was in motion. The M4 Sherman was also equipped with gyrostabilization.

The hull was modified, with the plates of the upper hull now constructed with welds instead of rivets. A single rounded, rolled plate now formed the rear engine deck, which simplified construction. A change known as the integrated fighting compartment reshaped the hull floor ammunition storage to be accessible from the new turret basket. The outdated sponson machine guns were finally dropped.

Production ceased in February 1943, with a total of 4,621 vehicles manufactured. The M3A1 served as an important training vehicle and featured prominently in the North African and Pacific Campaigns. The 1,594 M3A1s that went to the British were used as reconnaissance tanks in northwestern Europe. The Soviet Union received 340.

The M3A1 represented an effort by US Army Ordnance to improve the ergonomics and operation of the turret as well as modernize the fire-control system, a concept called the "integrated fighting compartment." The turret, type D58101, now would have the Combination Gun Mount M23, equipped with a Westinghouse vertical-axis gyrostabilizer. The turret received an Oilgear power traverse. And since having a power traverse, with a rate of rotation of fifteen seconds for 360 degrees, posed a serious threat to the crew when they had to scramble over the drive-shaft tunnel when operating the 37 mm gun, a turret basket now became standard equipment. The prototype M3A1, registration number W-306923, seen at Aberdeen Proving Ground on May 21, 1942, had the new D58101 turret, 37 mm Gun M6 on the Combination Gun Mount M23, gyrostabilizer, and power traverse. While production M3A1s would have a welded hull, the prototype was based on a riveted hull.

The prototype M3A1, registration number W-306923, is viewed from the left on May 21, 1942. The three pistol ports on the D58101 turret were positioned higher than those on the previous D39273 turret. Previously, the bottoms of the pistol ports were even with the top of the flared lower part of the turret.

As seen on the prototype M3A1, the mount for the antiaircraft machine gun on the D58101 turret was moved to the right side, and the bracket for the radio antenna was repositioned on the left side of the turret.

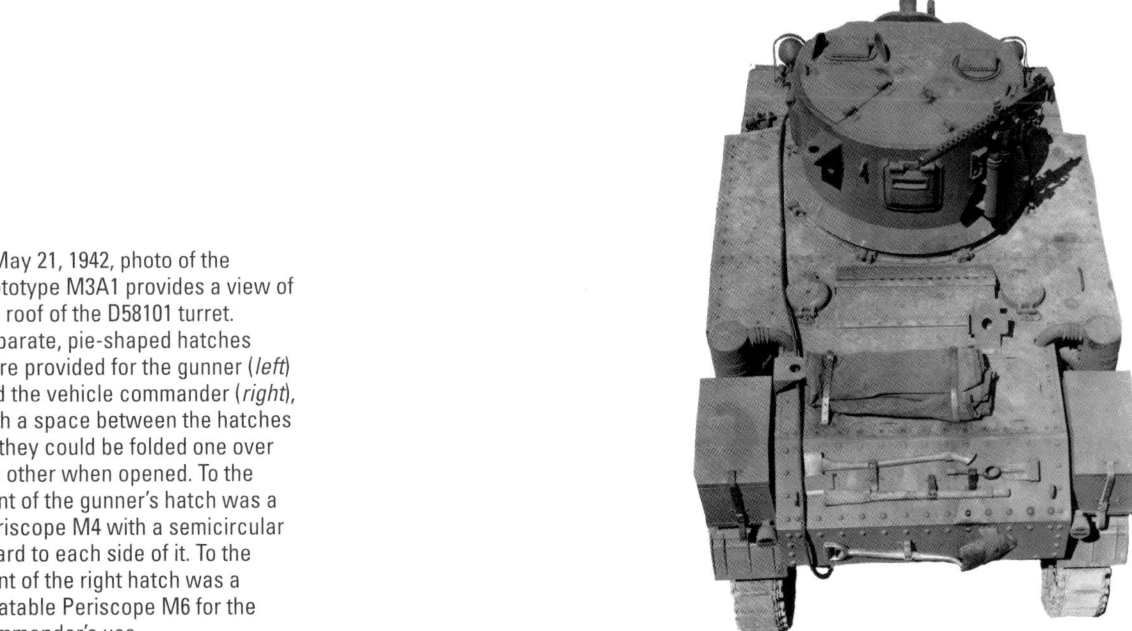

A May 21, 1942, photo of the prototype M3A1 provides a view of the roof of the D58101 turret. Separate, pie-shaped hatches were provided for the gunner (*left*) and the vehicle commander (*right*), with a space between the hatches so they could be folded one over the other when opened. To the front of the gunner's hatch was a Periscope M4 with a semicircular guard to each side of it. To the front of the right hatch was a rotatable Periscope M6 for the commander's use.

As seen in a front-overhead view of the prototype M3A1, the gunner's hatch door had a bent front edge, while the commander's had a straight front edge. Strapped to the left front fender is a machine gun tripod, which enabled the crew to set up the antiaircraft machine gun on the ground if required.

Whereas the flat, frontal plate of the turret had been attached with slotted, oval-headed screws to preceding models of the Light Tank M3, the M3A1 prototype and production vehicles had that plate welded to the sides of the turret. There were three lifting eyes on the D58101 turret: the two front ones are visible on the sides of the turret in this photo, and the other one was on the rear of the turret.

In a right-side view of the prototype M3A1, the position of the right-front lifting eye is apparent, with its shadow visible below it.

The rear lifting eye of the turret is to the left of the upper part of the rear vision port, as seen in another May 21, 1942, photo of the prototype M3A1 at Aberdeen Proving Ground. Note the design of the stiffeners, both fore and aft and lateral, built into the tops of the mudguards. Compare them to the plain, easily damaged mudguards in photos of earlier-production M3s. The two lifting eyes on the sloping rear extension of the engine deck, to the sides of the pioneer tools, were standard equipment on M3s.

Formerly in the collections of the US Army Ordnance Museum at Aberdeen Proving Ground, Maryland, was this Stuart Hybrid. Details of the turret and hull roof are displayed in this image, including the weld beads of the segments composing the flared bottom of the turret; the pistol port and its beveled splash guard, which is formed of weld beads; the right lifting eye; and the gun shield. Three of the eight roller and bearing brackets are in view: one below the lower front corner of the turret, and two alongside the turret ring. *Photo by author*

On the rear of the Stuart Hybrid turret are the antenna bracket, lifting eye, pistol port, and machine gun pedestal. In the foreground are the other radio antenna bracket, the engine's air intake hood and grille, and the right fuel-filler cover. The notch on the flared bottom of the turret represents a missing access plate, for servicing and cleaning the turret ring and rollers. *Photo by author*

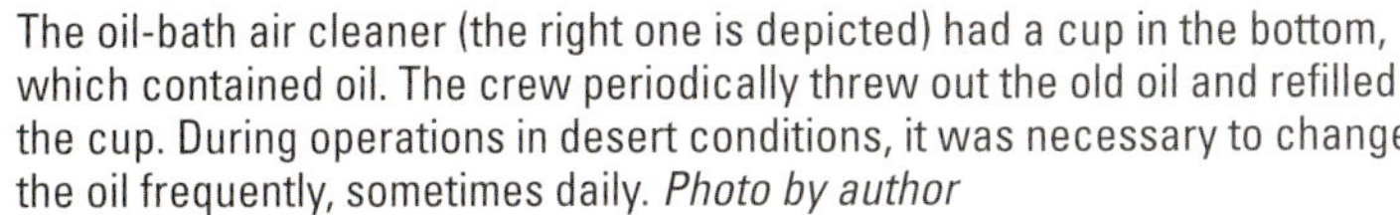
The oil-bath air cleaner (the right one is depicted) had a cup in the bottom, which contained oil. The crew periodically threw out the old oil and refilled the cup. During operations in desert conditions, it was necessary to change the oil frequently, sometimes daily. *Photo by author*

The sheet-metal storage box on the right-rear fender is viewed from the side. On the rear of the upper hull to the rear of that box is mounted the right taillight assembly. *Photo by author*

The right-rear corner of the Stuart Hybrid formerly at Aberdeen Proving Ground is viewed close-up. A latch and the three footman loops for a strap to hold the lid of the box closed remain in place. The tank has the side skirts favored by the British. *Photo by author*

A view from next to the engine access doors on the rear of the hull also takes in the idler adjustment mechanism, below the side panel of the mudguard. On the bottom of the hull is the right-side tow eye. *Photo by author*

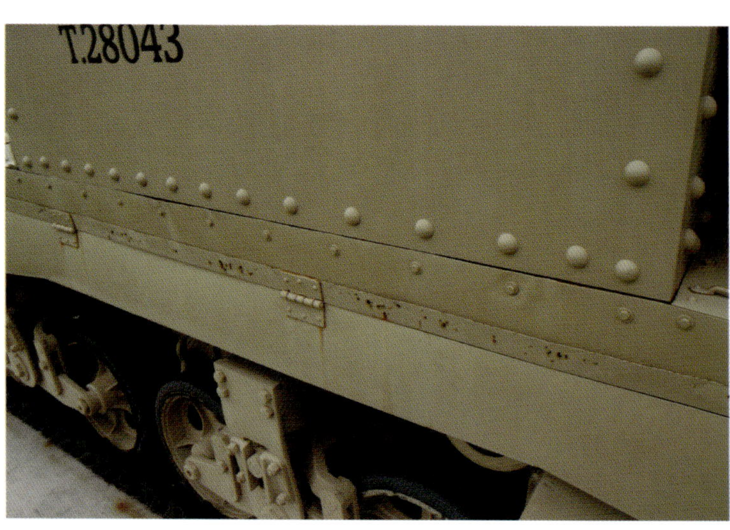

The middle part of the right skirt and parts of the sponson and the running gear are displayed. The skirt was hung on leaf hinges. *Photo by author*

The trailing-idler assemblies (the left one is depicted) were designed to give additional support to the rear of the vehicle and to provide an adjustable mechanism to maintain proper track tension. The track tension was adjusted by loosening the large nut over the wheel spindle and turning the small nut on the threaded shaft at the rear of the idler arm, to move the idler forward or backward. *Photo by author*

The front mudguard and skirt partially enclose the left sprocket assembly. Note the grease fittings on the side of the hub of the bogie wheel. *Photo by author*

The rough-cast texture of the gun shield and the cover for the bow machine gun contrast with the smooth surface of the glacis. Running across the glacis are several splash guards, designed to deflect small-caliber projectiles that strike the glacis. The box mounted on the right fender was not installed at the factory. The exposed electrical cables for the headlights pass into the hull through holes in the glacis. *Photo by author*

Photographed during its acceptance trials at Aberdeen Proving Ground on October 28, 1942, Light Tank M3A1, serial number 5260, exhibits the defining external characteristics of that model: the D58101 turret and the welded hull. Inside the tank were several new features of the integrated fighting compartment, including the Combination Gun Mount M23, the Oilgear power turret traverse, Westinghouse gyrostabilizer, turret basket, revised 37 mm ammunition storage racks, and an intercom system. This M3A1 has the long air hoses from the air cleaners to the engine compartment, a feature associated with vehicles equipped with the Guiberson Diesel engine.

The driver's hatch of an M3A1 to the left and the top of the photo frames a view of the new turret basket, which is traversed to about the 10 o'clock position. The gunner's seat is to the right, and the commander's is partially visible more to the left. Five ready rounds of 37 mm ammunition are stowed between the seats. Parts of the new 37 mm ammunition racks are visible at the bottom of the photo. Despite the open design of the turret basket, it was exceedingly difficult for the driver and assistant driver to enter or exit the tank through the turret.

A Light Tank M3A1 with gasoline engine is being subjected to a cold-weather starting test at the Tank-Automotive Center, Development Branch, Detroit, Michigan, on January 1, 1944. Icicles are hanging from the bottom of the curved rear end of the upper hull: this curved design was new for the M3A1. Marked on the forward end of the sponson is "ORD TEST VEHICLE" over "M3 355." To the right is a Light Tank M3A3. *TACOM LCMC History Office*

Light Tank M3A1, serial number 5260, is seen from above during acceptance trials at Aberdeen Proving Ground on October 28, 1942. The long air hose from the left air cleaner is seen from another angle. When the M3A1 and other M3s were equipped with Diesel engines, the air hoses were routed into the engine compartment through openings through the rear corners of the engine's air intake hood.

A clear view of the redesigned, curved rear of the upper hull is available in this image of a Light Tank M3A1 equipped with jettisonable auxiliary fuel tanks. Two straps are attached to each tank for lifting purposes. Stenciled on the rear of the tanks is manufacturer's data (US Rubber Co., Fisk Plant, Chicopee Falls, Massachusetts), serial and part numbers, date of manufacture ("8-42"), and "LEFT" and "RIGHT." *Central Susquehanna Community Foundation*

The late armor collector Fred Ropkey owned this Light Tank M3A1, with the nickname "El Diablo" painted in white script on the sponson above the registration number, W-307650. A detachable windshield is installed in the assistant driver's hatch. On the left front fender is a machine gun tripod with a canvas cover over the upper part of the tripod. *Skip Warvel*

Light Tank M3A1 W-307650 is armed with a water-cooled .30-caliber machine gun on the mount on the rear of the turret. A water-cooled machine gun in that position would have been an oddity, but not unheard of, on a Stuart during World War II. Even .50-caliber machine guns sometimes were installed on the turret mounts. *Skip Warvel*

An M3A1 owned and restored by Steve Greenberg, of Oregon, is equipped with jettisonable auxiliary fuel tanks to the sides of the turret. When these tanks were installed, they limited the traverse of the 37 mm gun to 180 degrees. Six grousers are stored on the side of the sponson. *Steve Greenberg*

The same M3A1 shown in the preceding photo, registration number W-3025312, bears the nickname “Katie Sue” on the upper forward corner of the sponson. Affixed to the side of the turret, to the front of the recognition star, is a placard with the crossed-sabers symbol of the US Cavalry. *Steve Greenberg*

A close-up from the right front of the M3A1 formerly in the Fred Ropkey collection shows the steel plug over the opening formerly used for the right-sponson machine gun (these guns were discontinued in the M3A1). At the center is the assistant driver's side vision slot, and to the right is a detachable windshield with a bare-metal frame. To the far right is the outboard hold-open strut for the assistant driver's armored hatch door. *Photo by author*

The turret of the Greenberg M3A1 is viewed from the right front. The sleeve over the 37 mm gun barrel to the front of the gun shield is the slide: a cylinder that provided support for the exposed part of the barrel and formed a structure for the barrel to slide in during recoil and recuperation. *Photo by author*

The driver's lower hatch, which was cut into the glacis, was equipped with a door consisting of two panels of armor plate with two leaf hinges joining them. The door is shown in the open position. The reason for the two-panel, hinged design was to allow the door to swing past the driver's front hatch door when it was raised in the open position. The procedure was for the driver to close the hatch door in the glacis before closing the front hatch door. *Photo by author*

A side view of the front of the left side of the turret illustrates the weld beads where the front of the turret joins the side, and where the flared lower part of the turret is attached to the side. To the side of the 37 mm gun is the coaxial .30-caliber machine gun with a perforated cooling jacket. *Photo by author*

The right pistol port on the turret of the Greenberg M3A1 fits snugly within a splash guard formed from a steel rim, with a beveled weld bead around it. *Photo by author*

A photo of the roof of the Steve Greenberg M3A1 turret provides a clear view of the weld beads between the roof and the left side and the front of the turret. The two hatch doors are open, the gunner's hatch lying over the commander's. Inside the turret, on the near outer wall is part of the mechanism for operating the left pistol port. Also inside may be seen the gunner's seat, a box for thermite smoke grenades, and the traversing gear box and the traversing motor. *Steve Greenberg*

The left jettisonable fuel tank of the Greenberg M3A1 is displayed, as are grousers stored on the side of the sponson and, behind the knapsack, the upper parts of the pistol port and turret. *Photo by author*

The left jettisonable fuel tank is shown, with the left air cleaner in the foreground. Webbing straps with metal fittings on the ends secure the fuel tank. *Photo by author*

Grousers, metal bars that were attached to tracks to give them better traction on ice and snow, are stored in racks made of angle iron on the sponson of the Greenberg M3A1. *Photo by author*

Clamp-type steel straps are holding the left air cleaner, manufactured by Norton, to the bracket on the rear face of the sponson of the Greenberg M3A1. The removable oil cup on the bottom of the air cleaner is secured with wing nuts. Note the weld seam where the side plate of the sponson abuts the rear plate. Also in view is the storage box on the rear fender. *Photo by author*

A view from the right rear of the Ropkey M3A1 shows the storage box on the right-rear fender, as well as the curved rear of the upper hull. The curved armor plate was secured with slotted oval-head screws countersunk into the armor. *Photo by author*

The antenna bracket on the engine deck of the Ropkey M3A1 is displayed, with an antenna installed. The front of the bracket is fastened to the right side of the hood of the engine's air intake, and the rear of the bracket is affixed to the deck. A screen has been fitted over the expanded-steel air intake grille on the engine, to the right. *Photo by author*

Underneath the overhang of the rear of the upper hull is the engine-air outlet grille, made of expanded-steel mesh. Above the grille are the exhausts and two mufflers. At the bottom of the photo is the rear of the lower hull and the engine access doors. *Photo by author*

Also visible in the preceding photo, this is a different type of antenna bracket mounted on the left side of the engine deck of a Light Tank M3A1, allowing the antenna to be held in a vertical position instead of at an angle. *Photo by author*

The turret basket of the Ropkey M3A1, which is traversed to about the 10 o'clock position, is seen through the driver's open hatch. The commander's seat is partially visible toward the left. On the floor of the turret basket are elements of the hydraulic motor, traverse hydraulic pump, oil pump, and slip ring. To the lower right are 37 mm ammunition racks. *Photo by author*

Stored upside down in the left ammunition rack are 37 mm ammunition rounds, with the projectiles color-coded as to their type. A flashlight is in a storage clip to the front of the ammo rack. In the left sponson, above the flashlight, is stored a first-aid kit. *Photo by author*

In a view through the driver's hatch of Steve Greenberg's tank, the turret basket is at the upper left, while at the bottom are, left to right, the drive-shaft tunnel, gearshift lever, fire extinguisher, driver's seat, and lubrication chart in its holder. In the rear corner of the compartment is the left 37 mm ammunition rack. *Photo by author*

The upholstery on the driver's seat is well worn. The white, finned structure at the lower left is the Synchromesh transmission, with five forward gears and one reverse. On the wall of the lower hull to the right are the lubrication chart and data plates. *Photo by author*

In a view down into the driver's compartment of an M3A1, the driver's seat and seat belts are to the top, to the front of which are the steering levers. To the lower right is a storage box for protectoscopes. *Photo by author*

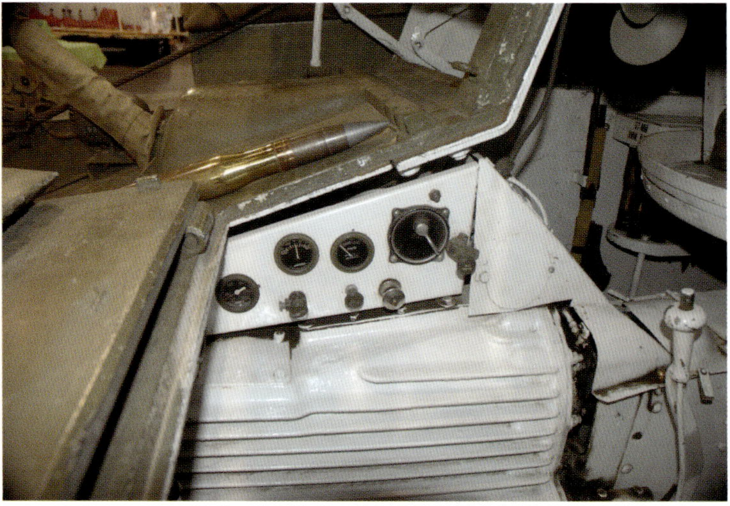

To the right of the driver's compartment is the Synchromesh five-speed transmission, on top of which is an instrument panel, which contains, left to right, an engine-oil-temperature gauge, ammeter, voltmeter, magneto switch, and wiper switch, as well as switches for lights, dimmer, and primer on the lower tier. A 37 mm round is lying on the glacis. *Photo by author*

In the driver's compartment to the front of the steering levers is the main instrument panel, which includes, left to right, the transmission-oil pressure gauge, engine hour meter, several switches and a hand throttle, and engine-oil-pressure gauge. Below the main instrument panel is the Pioneer compass. Below the compass are the clutch pedal, left, and the accelerator pedal, right. *Photo by author*

This is a driver's-eye view to the right into the assistant driver's compartment, where the detachable windshield has been installed over the open hatch. On the outside wall to the right of the center of the photo is the side vision slot. At the bottom is the speedometer, with two steel-rod guards over it. *Photo by author*

A Thompson .45-caliber submachine gun is on a storage rack to the front of the right 37 mm ammunition storage rack. To the upper right is the floor of the turret basket. *Photo by author*

The driver's and assistant driver's hatch doors are viewed from below, in a position on the left side of the tank. The interiors of the doors were painted white, and the mechanisms for the protectoscopes and the protected vision slots are present. *Photo by author*

A radio set and its dynamotor are mounted in the right sponson of an M3A1. On the stanchion of the turret basket toward the right, in front of the commander's seat, is a storage bracket for a canteen. *Photo by author*

In a view into the front right opening of the turret basket of an M3A1, the turret occupies the top half of the photo. Above the thermite smoke-grenade box, just above the center of the photograph, is the rear pistol port and protectoscope. The commander's and gunner's seats are on stanchions, part of the mechanism for raising and lowering the seats. Between the seats is a unit containing the traversing gear box and gear box shifter, hand traversing crank, and hydraulic traversing motor. The bare-metal object at the bottom center is the slip ring, also called the collector ring, an apparatus by which electrical current for components in the turret is continuously transmitted from the hull to the turret. *Photo by author*

In a view through the driver's hatch looking at the turret, at the bottom are the turret basket and the commander's seat, and at the top are the elevating mechanism and handwheel. The object shaped similar to a throttle quadrant to the right is the hand traversing crank. *Photo by author*

A brass data plate affixed to the lower rear corner of the recoil guard identifies this as a "MOUNT, COMBINATION, GUN, M23" and also includes the mount's serial number. *Photo by author*

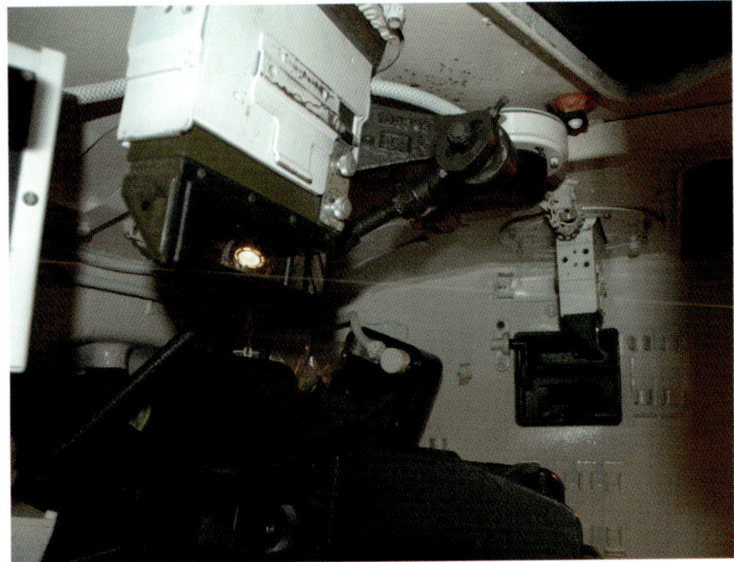

As seen from the left side of the turret interior of the Ropkey M3A1, in the foreground is the gunner's periscope, next to which is the Olive Drab–colored linkage that synchronizes the angle of the periscope with the elevation of the 37 mm cannon and coaxial machine gun. At the bottom are the gunner's telescopic sight and the breech and guard of the 37 mm cannon. The black box with the light-colored electrical cable attached to it on the far side of the cannon breech is the gyro control. On the opposite side of the turret are the commander's front periscope and the right pistol port and protectoscope. *Photo by author*

The guard on the left side of the Combination Gun Mount M23 in the Ropkey M3A1 protected the gunner's head and left shoulder from the recoil of the gun. The holes on the upper half of the guard were for fastening a cushion. Missing from this gun mount is the breech guard, which extended to the rear and included a spent-casing deflector. To the far left is the eyepiece of the gunner's Telescope M40. *Photo by author*

CHAPTER 3

Light Tank M3A3

The Light Tank M3A3 was the result of a request by the Armored Force for a revised version of the M3A1, with a new turret with a bustle and improved hull armor. The glacis of the M3A3 was similar to that of the Light Tank M5, but the M3A3 had sloped side armor. The lower hull, drivetrain, and running gear remained the same as for the M3A1. American Car and Foundry completed 3,427 M3A3s. The M3A3 in this photo is a late-production example, with "M3A3" and the number "13497" painted on the sponson and the glacis.

An interesting redesign of the M3A1 resulted in the Light Tank M3A3. The hull of the M5 Light Tank had inspired the Armored Force to request a similar hull for the M3A1. A very modern design emerged that featured a well-sloped glacis with side armor that angled 20 degrees from vertical. The forward hull roof now featured hatches for the driver and assistant driver. The new design could accommodate the installation of larger fuel tanks, negating the need for external fuel tanks and finally solving the long-standing range issue. Another innovation was the addition of a large sheet-metal bin mounted to the rear of the upper hull for personal crew stowage.

A bustle-like projection on the rear of the new turret allowed the radio to be mounted there or within the turret itself. A new main gun mount that increased accuracy was added on the basis of British data. Larger hatches facilitated turret egress.

The Light Tank M3A3 used the same drivetrain and suspension as that of the M3A1. A reduction of the final drive gear ratio was required to retain performance and reliability, due to an almost 4,000-pound increase in weight.

Between September 1942 and September 1943, American Car and Foundry produced 3,427 vehicles. Oddly, in spite of its many clever innovations, the M3A3 was produced exclusively for foreign aid and did not serve with US units. The largest users were the British, who received 2,045 units. The Chinese received 1,000, and these tanks are often seen in photos serving alongside the M4A4 model of the Sherman. Another 277 went to the Free French. Because of the ineffectiveness of the 37 mm main gun, some Commonwealth troops removed the turret to create a reconnaissance vehicle, known as the "Stuart Recce."

Light Tank M3A3, registration number W-3025643, is observed from the left side. Substantial skirts were installed on the vehicle; a hinged access door is present on the rear section of the skirt. Grousers are stored on the side of the turret, and two spare track links are stored on the rear of the upper hull. *Central Susquehanna Community Foundation*

A foul-weather hood with a windshield and wiper is installed over the driver's hatch of an M3A3. The glacis was 1⅛ inches thick, thinner than the 1½-inch frontal armor of the M3 and the M3A1, but the glacis was sloped at 48 degrees from vertical, which provided benefits in terms of improved protection and more space for the crew. The sides of the sponson were angled 20 degrees from vertical.

A stowage bin made of a metal frame and expanded-steel mesh is mounted on the rear of the upper hull of an M3A3. The storage bin is attached to the rear of a large storage box, designated the blanket-roll box, the top of which is higher than the top of the bin. The rear armor of the lower hull was of riveted construction. The radio antenna was mounted on an access panel on the rear of the turret. On the turret roof to the right of the antenna is a spotlight.

Light Tank M3A3 registration number W-3025643 is viewed from the upper left. A dust cover is over the 37 mm gun barrel. The size and shape of the storage box on the rear of the upper hull are apparent. The engine deck was a total departure from that of the M3 and M3A1, with four fuel fillers with armored covers, a new air intake grille for the engine, and new storage arrangement for the pioneer tools. *Central Susquehanna Community Foundation*

An M3A3 is viewed from the upper rear. On the rear of the storage box on the rear of the upper hull is a stencil identifying the storage location for a tarpaulin and a camouflage net in the storage bin. A stencil marking the location for a stored machine gun tripod is on the access panel on the rear of the turret; to each side of that plate are two footman loops for retainer straps for the tripod.

In an overhead view of an M3A3 may be seen the shapes of the two hatch doors and the locations of the periscopes: two rotating ones for the commander, on his hatch and to the front of his hatch, and one to the front of the gunner's hatch. On the engine deck are over twenty footman loops, for securing pioneer tools and other gear. The hatch doors for the driver and the assistant driver were hinged on their outboard rear corners, and each door had a rotating periscope.

This rare surviving Light Tank M3A3 is one of two on outdoor display in Richmond, Indiana. Grouser racks are welded to the side of the turret, and the pedestal mount for the machine gun has been removed from the side of the turret: the screw holes for the mount are visible to the front of the grouser racks. *Photo by author*

The brush guards for the headlights, which are fragile, remain in place on the glacis of this M3A3. The original ball mount for the bow machine gun is missing, and crude facsimiles have been installed. Note the heavy-duty splash guard running across the glacis and jogging down below the bow machine gun. The glacis actually constituted two plates of armor, with the joint between them being a little over an inch above the top of the splash guard. The lower plate was removable, for accessing the transmission and final drive. *Photo by author*

The front of the turret, the 37 mm gun barrel and gun shield, and the driver's and assistant driver's hatch doors are in view. Brush guards made of welded steel rods are screwed to the rotating mounts for the periscopes on these doors. Between the hatch doors is a ventilator hood, enclosed within a circular splash guard. The aperture for the gunner's telescope, seen here with a hood over it, was higher on the M3A3 than on the Light Tank M3. This was to give the gunner a better field of vision over the 37 mm gun barrel. The gunner's sight was the Telescope M54, part of the new Combination Gun Mount M44. *Photo by author*

In a view of the left front corner of the upper hull, the two track-like fittings on the upper part of the glacis to the front of the driver's hatch were standard features on the M3A3 and were associated with the driver's foul-weather hood. These fittings did not carry over to the subsequent Light Tanks M5 and M5A1. Centered just below these fittings is the driver's peephole, equipped with a steel plug and a splash guard fabricated from a U-shaped strip of steel and built-up weld beads. *Photo by author*

The left front fender/mudguard, the lower left corner of the glacis, and the brush guard for the left headlight and the siren are in view. Note the grab handle and the recessed hex-headed screws at the bottom of the glacis. *Photo by author*

The driver's hatch door with the rotating periscope mount and brush guard, part of the 37 mm gun shield, and the left half of the glacis are viewed from above the front of the turret roof. *Photo by author*

A view from above the right side of the turret includes the assistant driver's hatch door and periscope, as well as the fake barrel of the bow machine gun. *Photo by author*

In a photograph of the turret of an M3A3 displayed in Richmond, Indiana, note the flattened area on the 37 mm gun shield on the outboard side of the aperture for the coaxial machine gun. A similar flattened area is on the left side of the shield, and this was a standard feature on M3A3s. *Photo by author*

The bogie wheels and idlers on this M3A3 at Richmond, Indiana, are the type with steel plugs welded on the insides, to cover the openings between the spokes. This expedient made for stronger wheels. *Photo by author*

The right side of the turret of the M3A3 is shown in detail, including the two lifting eyes, the grouser racks, the screw holes for the missing machine gun mount, and, welded to the side of the turret above the grouser racks, a stop for the commander's hatch door. *Photo by author*

A prominent weld bead joins the front of the turret roof to the frontal plate of the turret, as seen from above. On the 37 mm gun shield, the manner in which the hood over the gunner's sight aperture where it melds into the swell above the gun barrel is visible. *Photo by author*

The engine's air intake grille on the M3A3 was a durable design, fabricated from steel rods on a steel frame with a grab handle on each side. At the top is the lower rear of the turret bustle. *Photo by author*

As seen from above the turret bustle of a Light Tank M3A3, on the gunner's (*left*) and commander's (*right*) hatch doors, to the fronts of the outboard hinges, are latches. When the doors are open, these latches engage with the stops projecting above the sides of the turret, to hold the hatch doors securely in place. The rotating periscope on the commander's hatch door has its hinged cover intact; the hinged covers on the other two periscopes are missing. *Photo by author*

The engine deck is observed from above the rear of the turret (*lower left*) and the radio antenna bracket (*bottom center*). It abounds with brackets and footman loops for securing pioneer tools. Parts of the two covers for the right-hand fuel fillers are to the lower left. On the sloped armored plate to the rear of the engine deck are the right taillight and two lifting eyes. Two brackets are mounted to the top edge of that sloped plate: they are two of the three brackets that the blanket-roll box was attached to. *Photo by author*

CHAPTER 4

Field Use

	Specifications			
	M3 early	**M3 late**	**M3A1**	**M3A3**
Weight	28,000 lbs.	28,000 lbs.	28,500 lbs.	32,400 lbs.
Length	178.4 in.	178.4 in.	178.4 in.	197.9 in.
Width	88 inches	88 in.	88 in.	99.4 in.
Height	104 inches	94 in.	94 in.	101 in.
Tread	73 inches	73 in.	73 in.	73 in.
Crew	4	4	4	4
Maximum speed	36 mph	36 mph	36 mph	31 mph
Fuel capacity	54 gallons	54 gallons	54 gallons	110 gallons
Range	70 miles	70 miles	70 miles	135 miles
Electrical	12 volts negative ground	12 volts negative ground	12 volts negative ground	12 volts negative ground
Transmission Speeds	5 Forward / 1 Reverse	5 Forward / 1 Reverse	5 Forward / 1 Reverse	5 Forward / 1 Reverse
Turning radius	42 ft.	42 ft.	42 ft.	42 ft.
Armament				
Main	37 mm M5	37 mm M6	37 mm M6	37 mm M6
Secondary	4 x .30-caliber	2 x .30-caliber	2 x .30-caliber	2 x .30-caliber
Flexible	1 x .30-caliber	1 x .30-caliber	1 x .30-caliber	1 x .30-caliber
Engine Data				
Engine make/model	Continental W-670-9A			
Number of cylinders	7			
Cubic-inch displacement	668			
Horsepower	262			
Torque	590			
Radio Equipment				
M3 vehicles were equipped with SCR-210 radios and RC-61 interphones. Command vehicles had an SCR-245.				
M3A1 vehicles were provided with SCR-508 radios with integral interphone. Command tanks had the SCR-506.				
M3A3 vehicles were equipped with either SCR-508, -528, or -538 radios, all with integral interphones. Again, the SCR-506 was fitted to command tanks.				

The vehicle commander of a Light Tank M2A4, registration number W-30493, stands in his hatch as the tank pauses along a road outside Mount Carmel, Louisiana, during 2nd and 3rd Army maneuvers in September 1941. On the rear of the turret is an Army Air Corps–style insignia: a white star with a red circle in the center, over a blue circle. A makeshift wooden rack on the rear of the upper hull holds several 5-gallon liquid containers.

The past meets the present as horse soldiers march past a Light Tank M2A4 from Company A, 67th Armored Regiment, 2nd Armored Division, during the Louisiana Maneuvers in September 1941. The white, trapezoidal shapes to the sides of the driver's and assistant driver's front hatches are the opened side panels, which on the M2A4 were fitted with hinges. The crewman to the rear of the turret is manning the .30-caliber antiaircraft machine gun.

Maj. Gen. Richard C. Moore, deputy chief of staff, US Army, is in the driver's seat of an early-production Light Tank M3 of the 2nd Armored Division at Ft. Benning, Georgia, on October 18, 1941. Note the inboard support strut of the driver's hatch door, which is folded up and secured with a spring clip on the door, and also the Air Corps–style star on the column between the driver's and assistant driver's hatches.

Tankers of the 2nd Armored Division are at their stations in an early Light Tank M3 with a D38976 welded turret at Ft. Benning, Georgia, on December 18, 1941. The early M3s lacked a turret basket, and the legs of the gunner standing in the cupola hatch and those of the vehicle commander seated in the left side of the turret are visible behind the driver and the assistant driver.

An early-production Stuart Mk. I with the D38976 welded turret is being unloaded from a ship at a port in the Middle East. This was one of many Stuarts transferred to the British and Commonwealth forces under the Lend-Lease program. A British Army registration number is on the sponson: T.27989.

A Stuart Mk. I stopped alongside a destroyed building in North Africa has received a number of British improvements, including two supports for the cupola hatch door, storage boxes on the front and rear fenders, smoke dischargers on the side of the turret, sand skirts, and rails to the sides of the sand skirts for attaching “sunshield” deception equipment. The sunshield equipment was designed to camouflage the tank to make it appear at a distance to be a lorry.

This Stuart Mk. I at Ft. Msus, Libya, is viewed close-up, showing its D38976 turret and the support for the cupola hatch door. The purpose of the object jutting from the cupola above the vision slot is not clear, but this feature was present on other British Stuarts in North Africa.

A column of Light Tanks M3 from the 10th Light Tank Company move out on a training exercise at Baldurshagi, Iceland, on May 3, 1942. This unit was formerly Company C, 70th Tank Battalion, until it was detached and assigned to garrison duty in Iceland in February 1942. All four M3s have the D39273 turrets.

The same column of M3s from the 10th Light Tank Company is viewed from the left side during a training exercise on a tundra in Iceland on May 3, 1942. Vehicle numbers are painted on the turrets: the first three are numbered 16, 4, and 9. At least the first two tanks have a machine gun tripod stored both on the left and the right front fenders.

Maj. Gen. Alexander Patch is standing with a tanker on the engine deck of Light Tank M3, registration number W-302714, during an inspection of armor of the Americal Division on New Caledonia on May 31, 1942. The number "5" on a roundel is placed at two spots on the right side of the D39273 turret.

The initialism "AFS," for Armored Forces School, is painted on the recognition star on this Light Tank M3 with a D38976 welded turret, maneuvering over rough, brushy terrain during an exercise at Ft. Knox, Kentucky, in June 1942. *Library of Congress*

Light Tank M3 registration number W-301471, from the Armored Forces School at Ft. Knox, is advancing through a woodland during a training session in June 1942. *Library of Congress*

A signal flag is jammed into the pedestal mount for the antiaircraft machine gun on the side of the D38976 turret of a Light Tank M3 during an Armored Forces School training exercise at Ft. Knox in June 1942. *Library of Congress*

A recognition star has been painted on the inner side of the cupola hatch door of an Armored Forces School M3, seen during a training operation in June 1942. *Library of Congress*

A crewman peeps over the top of the cupola of a Light Tank M3 during a training exercise at the Armored Forces School in June 1942. Note how the light-colored band painted on the turret adheres to the curve of the top of the gun shield. *Library of Congress*

Two of the Armored Forces School's Light Tanks M3 are advancing along a dirt road at Ft. Knox, Kentucky, in June 1942. The bow and sponson machine guns have been removed from these vehicles. The men standing in the cupolas are wearing an early style of tanker's helmet. *Library of Congress*

In an original color photograph taken during a training exercise at Ft. Knox in or around June 1942, two M3s exhibit yellow stripes and recognition stars on their D38976 turrets. "AFS" is marked on the stars. *Library of Congress*

Douglas A-20 Havoc attack planes fly low in the background behind a formation of US Army M3s at the Desert Training Center, near Indio, California, in mid-1942. Most but not all of the tanks are marked "M" on both sides of the turret: this reflected a temporary designation of these particular M3s as medium tanks for purposes of the training program. Markings on the bows of the tanks are for Company E, 34th Armored Regiment. The Stuart at the front center and several others exhibit the long air intake pipes of the carburetor on the Diesel-engine-powered M3s.

The nickname "DEBUTE" is painted on the sponson of a Light Tank M3 with no tracks that is about to be towed by a Diamond T wrecker during the Tennessee Maneuvers in September 1942. The vehicle was assigned to the 4th Armored Division: the "84 R" in the unit markings on the bow appear to have pertained to the 84th Reconnaissance Troop.

A Stuart with the D39273 rolled, homogeneous-armor turret has paused outside Sidi Barrani during the advance of the British 8th Army on Tobruk, Libya, in November 1942, to receive fuel from a Royal Air Force tanker truck. Angle-iron brackets for grousers are on the side of the sponson. On the right front fender is a storage box of the model mounted on the rear fenders. Helmets are stored on the brush guards on the front fenders.

During a rehearsal of an invasion landing, a US Marine Corps Light Tank M3 advances up a beach in Australia in April 1943 while officers look on in the background. "USMC" is stenciled on the lower front portion of the side of the sponson; farther aft on the sponson is the USMC registration number, which appears to be 60062.

Two African American tankers of the 51st Composite Battalion, US Marine Corps, are standing in the turret hatches of a Light Tank M3A1 at Camp Lejeune, North Carolina, in March 1943. Jettisonable auxiliary fuel tanks are mounted on this tank, and a dust cover is over the 37 mm cannon barrel and gun shield.

Light Tank M3, registration number W-306582, is sitting in a muddy equipment yard at a facility referred to in the original label of the photo as "MBS Ordnance," in Oran, Algeria, on April 5, 1943. The "MBS" initialism stood for Mediterranean Base Section. The machine guns have been dismounted from the sponson and the bow. Another Light Tank M3 is in the right background.

The same M3 shown in the preceding photo, W-306582, is viewed from the right rear at Mediterranean Base Section Ordnance, Oran, Algeria, on April 5, 1942. Note the damage to the right taillight assembly and to the rear of the tray that supported the storage box on the right rear fender.

A crewman dressed in overalls and wearing an M1 helmet sits on the cupola of a Light Tank M3 with D39273 turret in the Tunisian desert in April 1943. Grousers are stored on the front of the glacis, and a bedroll is stashed on the turret roof. The brush guards are missing from the front fenders, and the headlights are turned downward.

A USMC Light Tank M3, registration number 60126, advances to support infantry during a rehearsal for an invasion in Australia in April 1943. A lighter color of paint has been painted in swatches over the base color of the turret and upper hull, for camouflage.

In a photograph apparently taken at the same time and place as the preceding one, a Marine Corps Light Tank M3 with the nickname "Eight Ball" painted on the D38976 turret is advancing along a dusty road in Australia in April 1943. A small number "4" is painted on the turret, below the nickname.

The USMC Light Tank M3 nicknamed "Eight Ball" and numbered "4" is driving on mesh matting on a sandy beach in Australia during a rehearsal for an invasion landing of an unidentified island in the South Pacific in April 1943.

The Australian army obtained a number of Light Tanks M3 during World War II, such as this example from 2/8 Armoured Regiment, which is fording a stream outside Popondetta, Papua New Guinea, on May 19, 1943. A length of spare track links are stored, outer side up, on the glacis.

A US Army photographer took this image of an abandoned Light Tank M3 that the Germans captured and put to their own use. The photo is dated September 1943. In the distance is an ancient aqueduct.

On New Caledonia on October 10, 1943, a Light Tank M3A1 is preparing to tow a captured Japanese tank, in order to start the enemy tank. This tank is equipped with jettisonable fuel tanks and, in the bow machine gun mount, a flamethrower. Marston matting, steel planking used for constructing airfield runways, has been attached to the glacis and on the side of the tank for extra protection.

The same flamethrower-equipped M3A1 seen in the preceding photograph is demonstrating its firepower by firing unthickened fuel at a base on New Caledonia on October 10, 1943. The front fenders have been removed, possibly to allow a better field of fire and to prevent the sheet metal of these structures from being damaged by the flamethrower.

A Light Tank M3A1 armed with a flamethrower in the bow position, possibly the same tank shown in the two preceding photographs, is viewed from the left front in a field on New Caledonia on October 10, 1943.

On November 20, 1943, a crewman of a Light Tank M3A1 of Company C, 165th Infantry Regiment, that became bogged in a bomb crater on Butaritari, Makin Atoll, Gilbert Islands, is lying in water behind the tank, removing chunks of coral from the track before a recovery vehicle attempts to free it. On the engine deck is the lower part of an intake trunk, part of a deep-fording kit. Interestingly, the bag lashed to the top of the folded tarpaulin on the engine deck is a coffee-bean bag.

An M3A1 attached to Company C, 165th Infantry, advances from Red Beach at Buritari, Makin Atoll, into a palm grove on November 20, 1943. A battered deep-fording trunk is on the engine deck. Painted on the left side of the rear of the upper hull is a delta, the symbol of armor. On the right side is the letter *C*, indicating the company, over the number "43," the vehicle's number in its unit's order of march.

The majority of Light Tanks M3A3 were exported to various Allied nations during World War II. Here, Chinese tankers are receiving instructions on operating an M3A3 at Ramgarh, India, in late 1943. Stenciled on the glacis of each of these tanks are instructions on unhooking the tow cable before towing the tank. Starting in October 1943, Ramgahr was a training base for the 1st Provisional Tank Group, a Nationalist Chinese army unit with six tank battalions and an American commanding officer, Lt. Col. Rothwell Brown. A shipment of 145 Light Tanks M3A3 was the first type of tank the group received.

Light Tank M3A3, registration number 3027625, is racing across a dusty field at Ramgarh, India, in late 1943. The crewmen were members of US Army Ordnance: specifically, they probably were members of the 527th Ordnance Company, and they were putting the tank through a test run. Stenciled in white on the sponson in small characters is "SCR 538," a reference to the installation of an SCR-538 radio receiver in the vehicle. There was an HQ, the US Army's 527th Ordnance Co. (Heavy Maintenance) (Tank), attached to the 1st Provisional Tank Group.

During the invasion of the Philippines in 1942, the Japanese captured and put to use large numbers of American vehicles. Such was the case with this Light Tank M3 with a D38976 turret, found by the Americans when they liberated Manila in early 1945. The 37 mm gun and its shield were missing from the tank.

Members of the 145th Ordnance Motor Vehicle Assembly Company, US Army, are test-driving a Light Tank M3A3 in front of a long row of M3s at Victoria Park, Calcutta, India, on April 12, 1944. These tanks recently had arrived in India, for use in the China-Burma-India (CBI) theater. The tank being driven has remnants of sealant material on the 37 mm gun and its shield, while the parked tanks have the sealant materials intact, and their 37 mm guns are resting on wooden travel locks.

Chinese tankers from the 3rd Company, 1st Battalion, 1st Provisional Tank Group, are operating M3A3s in a muddy field outside Ramgarh, India, on March 4, 1944. A white diamond marking is on the upper center of the glacis of both of these M3A3s.

A Light Tank M3A1 armed with a British Ronson flamethrower in the turret is assaulting Japanese snipers who have infiltrated into a rear area on Saipan in July 1944. M3A1s thus armed were nicknamed the Satan. The part of the flamethrower that projected from the front of the turret was in the form of a short, thick tube mounted on a flat, round shield. The dark shape to the front of the turret that looks like a gun barrel in this photo actually is the stream of fuel issuing from the flamethrower.

Liberation day! Joyous residents of St.-Sauveur, France, greet the crew of a Light Tank M3A3 numbered "17" on the sponson and attached to a French armored unit in August 1944. Painted on near the rear of the sponson is a tricolor French flag.

A column of tanks from the British 34th Armoured Brigade, including a Stuart III (M3A1) in the front, is passing through a Dutch village in late 1944. A large, white recognition star inside a white circle is on the turret roof. Note that the thin piece of turret roof between the hatch doors has been removed, to make for an enlarged opening.

Following World War II, the United States retired its M3 and M5 light tanks, but for years after that conflict, these vehicles remained in use by various armies around the world. One such vehicle, an M3A3, is seen here supporting Indian troops at a roadblock on the Semarang-Ambarawa road in Java during the Indonesian War of Independence in December 1945.